Eggs All The Year Round
The Successful and Profitable Keeping of Poultry for Eggs and Meat

by James Maclehose

with an introduction by Jackson Chambers

This work contains material that was originally published in 1876.

This publication is within the Public Domain.

This edition is reprinted for educational purposes and in accordance with all applicable Federal Laws.

Introduction Copyright 2017 by Jackson Chambers

Self Reliance Books

Get more historic titles on animal and stock breeding, gardening and old fashioned skills by visiting us at:

http://selfreliancebooks.blogspot.com/

Introduction

I am pleased to present yet another title on Poultry.

The work is in the Public Domain and is re-printed here in accordance with Federal Laws.

As with all reprinted books of this age that are intended to perfectly reproduce the original edition, considerable pains and effort had to be undertaken to correct fading and sometimes outright damage to existing proofs of this title. At times, this task is quite monumental, requiring an almost total "rebuilding" of some pages from digital proofs of multiple copies. Despite this, imperfections still sometimes exist in the final proof and may detract from the visual appearance of the text.

I hope you enjoy reading this book as much as I enjoyed making it available to readers again.

Jackson Chambers

TO

MY AUNT,

J. C. M.,

WHOSE INTEREST IN POULTRY WAS AWAKENED BY
THE RESULT OF THE EXPERIENCE WHICH
SUGGESTED THESE PAGES,

*I GRATEFULLY AND AFFECTIONATELY
DEDICATE THIS LITTLE BOOK.*

CONTENTS.

	PAGE
PREFACE,	9

INTRODUCTION—General ignorance about poultry management—Opinions of Stephens and Mechi—In France poultry rearing an independent branch of industry, . . 13

Part I.

CHAPTER I.—THE HEN-HOUSE.

Height—Materials for building—Floor—Roof—Furniture—Covered shed—Yard—Portable houses—Lime washing—Nests—Feeding Dishes—Dust baths, ... 23

CHAPTER II.—THE BREEDS.

Sitting Breeds — Dorking—Game — Brahmas—Cochins — Non-sitting breeds—Hamburghs—Minorcas—Andalusians — Anconas — Columbians — Polands — Houdans—Crêve-Cœur—La Flêche—Leghorns, ... 29

CHAPTER III.—FEEDING OF POULTRY.

Danger of over-feeding—Selection of food—Hours of feeding—Ingredients of morning meal—Malt-meat—Green food — Grain — Indian corn — Wheat — Oats—Barley—Buck-wheat—Hemp-seed—Water—Lime water—Drinking vessels, ... 37

CHAPTER IV.—HATCHING AND REARING OF CHICKENS.

Incubators—Selection of mothers—Sitting pens—Hatching nests—Number of eggs to be set under each hen—Assisting weak chickens to remove from the shell—First food

CONTENTS.

of chickens—Way to feed by candle-light—Patented meals—Crushed bones—Quantity to be given—Advantages of glass fronts to coops during early months of the year—Wooden floors, ... 45

Part II.

CHAPTER I.—GENERAL SUMMARY.

Personal supervision—How the amateur should begin poultry-keeping—Merits of breeds—Purchasing stock-birds—Proper season to make up breeding yards—Proper seasons to set eggs—How to make hens sit at any time—How to prevent hens sitting—Cockerels—Pullets—Moulting—Proper age to kill hens—Diseases—Preserving eggs, 57

CHAPTER II.—GENERAL TREATMENT.

Number of hens in each breeding yard—Way to keep setting eggs—Feeding of sitting hens—Length of time hens should be absent from nest—Examination of nests—Way to discover fertile eggs—Watering the eggs—Removal of hen and chickens to coops—Feeding of hen and chickens—Artificial mothers—Capons—Way to train a capon—The art of caponing—Age when cockerels must be separated from pullets—Fattening cockerels—Sign of laying in pullets and hens, ... 68

CHAPTER III.—DAILY DUTIES.

Examination of hen-houses—Way to discover sick fowl—Punctuality in feeding—Morning meal—Number of laying hens to be kept in one yard—Warm food—Horse flesh—How to construct a "verminal"—Tallow-chandler's greaves—Seasoning the food—Mid-day meal—Evening meal—Mixing of food—Feeding dishes—*Quantity of food to be given at each meal*—Way to fatten fowls for table use—Way to kill a fowl, ... 80

CHAPTER IV.—PROFITS.

Eggs at fourpence per dozen—Chickens at fourpence per pound—Cotters' and farmers' profit—Value of poultry manure—How three hundred eggs per year may be obtained—Hens sold by the pound—Each hen capable of returning twelve shillings annual profit—Chickens for the market—Value of spring chickens, ... 87

PREFACE.

I PROPOSE in the following pages to show how poultry may be made, not only to pay, but to pay *well*—in short, to yield *enormous profits*. I wish to prove, that by careful and judicious management, eggs may be produced at a cost of *fourpence per dozen*, chickens at *fourpence per pound;* and that each hen kept, may be made to return a profit of ten shillings per annum. I do not pretend to say that this result can be attained *at once*, by any one who has no previous knowledge of the management of hens. In poultry-keeping, as in everything else, one can only arrive at the requisite knowledge after some experience: yet it is always well to have a friend who, himself thoroughly acquainted with the management of poultry, is willing to give sound advice, and to guide others to the attainment of the required information. I propose to act as that friend, and, if the reader will follow *faithfully* and *without reserve* the advices given, I can promise, that he will soon acquire a competent acquaintance with the hatching and rearing and keeping of fowls, together with every detail regarding the production of eggs. To many this information *may be*, as I fervently hope it *will be*, the means of their gaining a respectable livelihood, if only I can induce them to begin poultry-keeping as a separate branch of industry.

I have long thought of writing a treatise on the management of fowls, but have from time to time delayed, in the hope that some other would undertake a work which, I think, is one of the most necessary of our time. It is true that there are *many, very many* books dealing with this subject, in which any one, if he chooses to study them, may find all and a great deal more than is contained in this little book. I cheerfully accord my gratitude to the several authors of these works, for the way in which they have demonstrated that the long neglected hen may be made to pay. And I am not without hope, that the influence of their united teaching, (and I trust that through *this treatise*, I may succeed in aiding in some small measure,) may be felt throughout the length and the breadth of our land, and that very soon the miserable, deformed, sickly creature now called by the name of hen, will become a thing of the past, because giving way to purer and better breeds. But the great desideratum which I think exists, is *not* that there are *no books*, but that these books are both *expensive* and *voluminous*, and are consequently either not within the reach, or would take up too much of the time of those for whom I specially write. My desire is to supply that desideratum,—my only and great aim is to combine in as *brief* and *comprehensive* and *cheap* a form as *possible*, all the information that is essential for a thorough knowledge of, and practical acquaintance with, the successful keeping of poultry.

One other reason for putting into execution at this time a long considered, though often deferred plan, of disseminating *cheaply* some information about poultry, which might prove a handbook and guide for the million, is the increased price of every commodity. The natural consequence of this is, that not a few have great difficulty in making the two ends of their small incomes meet, and are compelled to undergo many hardships before they can do so. Now I think, something might be done—and it is with this object I am writing—to make the poor feel their poverty less severely, and those with limited means their hardships, by endeavouring to induce *all* who have the necessary accommodation, to undertake the rearing of chickens and the production of eggs. *Chickens*, and more especially *eggs*, are sold at such *high* and *exorbitant prices*, that the poorer classes are almost entirely precluded from purchasing them at all. And yet, if due attention were only given to the management of poultry, the market prices of both would become *very much reduced;* and, at the same time, very handsome profits would be immediately realised by the man who engaged, as a matter of business, in their production.

My treatise proposes to show *clearly* and *distinctly* how the supply may be made much more equal to the demand, —how *quickly* and *surely* large profits may be obtained by anyone, who is prepared to act up willingly and implicitly to the directions given. The first part deals entirely with the particulars of poultry management—the second part, with the application of these particulars to special and individual cases. To one who has no previous knowledge of the treatment of hens, the first part will be found of *great importance* in giving the required knowledge—the second part will shew, how all this knowledge may be so utilized, as to make poultry-keeping a *great success*. I therefore recommend, that after the first part has been read and understood, the second part be *mastered thoroughly in all its details.*

I must further premise, that my treatise is not intended for the mere poultry-fancier, who cares neither for production of eggs nor of flesh, but whose sole aim is to breed for points. I am writing simply for the benefit of those, who may wish to know *how* eggs and fowls can be profitably obtained. I am giving in these pages the result of long experience and much study; and if I succeed in inducing some to commence poultry-keeping as a separate branch of industry, and am thus in any one way of service, in aiding them to gain their livelihood, or to become independent in the world, or if, as a natural and necessary result, I may be the means, in however small a degree, of reducing the present high prices of both chickens and eggs, and of thus relieving the poverty of the poor, and diminishing the hardships of those with small incomes—I will have the reward I seek, in feeling that my labours have not been in vain.

INTRODUCTION.

THE universal belief is that it is impossible to make poultry pay; and certainly it is a very general experience that eggs are at present, when purchased, only obtained at most extravagant prices, or where hens are kept, at considerable loss. Now the wonder is, that with this experience so many would continue to keep hens. The great value of a hen is in the production of eggs, but if she costs more than she yields, she is for all practical purposes absolutely useless. I am glad, however, to be able to state, that if hens are unprofitable the fault lies in the general management.

The first great fact which will become at once obvious to any one who begins the study of hens, is, that *perhaps of all animals* they are the most neglected, because the least understood. Men keep them in the most careless and indifferent way, and with irregular feeding—some days with no feeding at all—with wretched, ill-ventilated, houses for their accommodation—with no attention paid to their comfort or cleanliness—expect that they should give for *this generous treatment*, an abundant supply of eggs. The absolute fanaticism of this expectation will become apparent, if one applies the same principle to other animals. If pigs are not fed regularly and liberally, they will *not* fatten at all; if cows are not supplied with food abundantly, they will not yield milk; and if poultry are insufficiently fed and cared for, they will *not give—nor can they be expected to give*—any profitable return in either flesh or eggs. It is simply because of determined neglect and gross ignorance, that poultry are not made, of all live stock, by far the most profitable and productive. Perhaps, the following quotation from Stephens' "*Book of the Farm*," may be not inappropriate to this part of my subject, written as it is by a man who has made fowls, as well as every other animal connected with a farm-yard, a *special study*, and who, as an *authority* on all these subjects, stands unrivalled. It emphatically corroborates all I have said:—

"Of all animals," he says, "reared on a farm, none are so much neglected by the farmer, both as regards selection of their kind and disposition to fatten, as every sort of domesticated fowl. Indeed, the supposition that any farmer should devote a *part* of *his* time to the consideration of poultry, seems to be regarded by him as an unpardonable affront on his manhood. Women only, in his estimation, are fit for such a charge; and doubtless they are the best, and would do it well, were they not begrudged of every particle of good food they may bestow upon poultry. The consequence is, as might be expected, the surprise of finding at any farmstead a single fowl of any description in *good* condition—that is, such as might be killed at the instant in a fit state for the table. The usual objection against feeding fowls is, that they do not pay, and no doubt the market price received for lean, stringy fleshed, sinewy-legged fowls is far from remunerative; but whose fault is it they are sent to market in that state, but the rearer of them? and why should purchasers give a high price for fowls in such a condition? Some excuse might be made for having lean fowls, were any difficulty experienced in feeding them, but there is none; and the idea of expense is a bugbear, and, like other fears, would vanish were fowls reared more in consonance with common sense than they are, on the notion they can never be ill off if at liberty to shift for themselves. Such a notion is founded on a grievous error, in rearing any kind of live stock on a farm. Better keep *no stock at all*, than rear them on such a principle. Fowls may be deemed a worthless stock, and they are so generally, but only on account of bad management. Apart from every consideration of profit, every farmer has it in his power, *at all times*, to have a well-fed fowl on his table; but he cannot have it while grudging the food required to make it so. He may rest assured that *good* poultry always at command would save him many a butcher's bill, which must be settled in cash—and *cash* the farmer has only by sale at market of some commodity of the farm. Few farmers kill their own mutton—that is, keep fat sheep for their own use; lamb they do kill in the season; but as to beef, it is always purchased, so that their families greatly depend upon the produce of the poultry-yard and pig-sty for their meat diet, and it is quite in their power to have these at all times in the highest perfection. . . . And it is no small luxury to enjoy a new-laid egg at breakfast every winter morning—a luxury which I enjoyed as many years as I lived in the country."

Again he says,

"I have often heard it expressed as a decided opinion, that it is impossible to feed fowls with a profit. It seems to me strange that fowls should not make a return for their keep when the other animals on a farm do; so I cannot coincide with the opinion until I have seen the experiments fairly tried by a farmer; and so far as my own limited experience instructs me, my opinion is in the opposite direction."

No unprejudiced reader but will at once admit, that the statements made by Mr. Stephens are well worthy of consideration.

As confirmatory of the foregoing, I should like to refer to the experience of Mr. Mechi of Tiptree Hall, whose exertions in the cause of agricultural progress have obtained for him far more than a mere local fame, and whose opinions on the management and profit of farm live-stock —of course including poultry—have most deservedly been considered *very valuable and important*. In his "*Farm Balance Sheets*" in answer to the question, Do poultry pay? he says,

"That is a very proper question to ask, for I hold as a principle that everything that does not pay, directly or indirectly, should be abandoned as commercially and agriculturally wrong. I test this by the following propositions

INTRODUCTION.

and comparisons :—It pays indirectly to produce cattle, sheep, and pigs—if so, why not poultry? It does not cost more to produce a pound of poultry, live weight, than a pound of meat live weight. Well then, do poultry sell at more or less per pound live weight than meat? The answer to this is, 'a pound of poultry live weight, always sells for much more than for a pound of meat live weight,' and so the question is settled in favour of poultry. The average price of first quality beef is five shillings per stone, or seven pence half penny per ℔. for the carcase net dead weight. The live weight price would therefore be 4½d. per ℔., because, according to the best authorities, fat beasts lose 40 per cent. of their live weight in offal. The average live weight price of the best poultry is quite 9d. per ℔., or double the price of beef, and not quite double the price of mutton. At this time of year poultry sell, wholesale, for much more than double the price of meat. Those I have sent to market recently have sold for 11½d. to 1s. per ℔. live weight—against 4½d. per ℔. live weight for beef, which is its present price. That settles the question so far, and the verdict is strongly in favour of poultry. Poultry are evidently dear food to the consumer; but does it cost more food to produce a pound of poultry than a pound of meat live weight? I answer *decidedly not*, but the reverse; for my cattle and sheep don't eat worms and insects, whereas fowls consume them abundantly, and economise and apply every scattered seed or kernel that would otherwise be wasted. In another point of view—is the cost of attendance and shelter greater with poultry than with cattle? I reply *No*.

As to the production of eggs, that depends upon the quality and quantity of food administered and the accompaniments of proper warmth and shelter. There is no fear of overstocking the market with either eggs or poultry. We consume daily one million of foreign eggs. I generally keep from three to four hundred fowls. They have free access to every field during the whole year, and although they help themselves at harvest time, when the corn is in sheaf, I always get my best crops of corn on fields adjoining the hen-house. I have this year (1867) two fields of wheat drilled, and only one bushel of seed per acre. They come within ten to twenty yards of the fowl-house, and are a perfect plant, although the poultry has been scratching and cultivating the fields ever since they were drilled. We are apt to forget that fowls, like sheep, manure where they go. I must say I used at one time to feel angry and nervous when I saw them hard at work on the newly sown corn, but I soon learned to feel confident that insects were their principal gain, and that my well and deeply-deposited corn escaped. . . . I look upon them now as the farmer's *best friends*. It is essential to the well-doing of poultry that they should have free access to grass, clover, and green food. They are very fond of mangel and swedes, but I never found that they injured them while growing. . . . Nothing pays better on a farm than a good stock of poultry properly managed. With them everything is turned to account; not a kernel, wild seed, or insect escapes their scrutinizing eyes. Their industrious claws are ever at work uncovering, ready for appropriation, every hidden but consumable substance. . . . They are very useful in clearing off flies. I have often been amused at seeing the neat and quick manner of their taking flies from reposing bullocks and sheep, much to their comfort."

These quotations from books, written by men whose experience on every subject upon which they write has been amply tested, will, I hope, aid in convincing the reader, that I am not without the best authority for stating, that poultry can be made to pay well.

If further proof be wanting, it may surely be found in the fact, that on the Continent poultry-keeping has become in a great measure an independent branch of industry; and that, besides providing amply for their own home consumption, the French find it most profitable to import *very largely* both fowls and eggs into Great Britain. More than five hundred millions of eggs, and many thousand tons of poultry are annually imported, and yet this vast quantity, together with what is produced in the country itself, falls *very very* far short of the demand.

It does almost seem a matter of profound wonder that *we*, who are generally considered one of the most far-seeing and shrewd nations in the world, should have allowed another to step ahead of us in this matter, and blindly shut our eyes to this truth, that by failing to cultivate poultry and eggs as a meat supply, we are positively putting into the pockets of foreigners, millions of money that might, to very great advantage, be circulated among our own labouring people. Now I think it must be admitted, that if French farmers can make poultry pay, we should be able to do the same. If poultry-keeping, as a separate branch of industry, is found in France to be profitable, it should most unquestionably be found so here. We have equal facilities, and, being on the spot, we can command an immediate market, and forestall the necessity for importation at all. Besides, even if we could not at once check importation, we need not fear it. The demand is so very much in excess of the supply, that, even were the supply doubled, we would, with an increasing population, be still unable to meet it. There is not the slightest doubt that any man of ordinary intelligence, if possessed of a small plot of ground, would make a respectable livelihood by cultivating more closely and carefully this branch of industry. One great advantage it offers to everyone is, that almost no capital is required, and that what little capital is employed, is being continually turned over, each time realizing a high percentage.

PART I.

CHAPTER I.

THE HEN-HOUSE.

It is essential to have a good hen-house—*light*, *airy*, and *free from damp*. It should have as much height as possible, so as to give a greater circulation to the air, and to prevent at any time (and more particularly during the long nights of winter) the atmosphere becoming tainted. Proper ventilation *must be* obtained without any approach to draught. What is known as the "*louver boards*," is the simplest and cheapest way of ventilating.

Fowl-houses may be constructed of wood without the fear of their being *too* cold, if only the boards are *tongued*, or have a spar nailed over each junction. This, when tarred, will effectually prevent all draught.

The floor must have a hard, and if possible a smooth surface, so that it can easily be swept clean each day. An asphalt floor is the best, the only objection being its expense. I would strongly recommend, as perhaps the cheapest, riddled ashes well mixed with hot lime. After mixing, it requires to stand for a few days, when it must be again turned with the spade, and its thickness reduced by adding water. It requires to be laid on the ground two or three inches thick, and to have some bullock's blood or tar poured over the surface. In a few days, if allowed to stand without being used, it will form a hard, smooth and excellent floor.

A window in the gable or roof, and if possible opening to the south or south-east, will admit sufficient light.

The roof may be slated or constructed of wood, covered with felt, and tarred.

The only furniture admissible in a fowl house are laying-boxes and roosts. The former should be placed in the darkest corner of the house, with a small aperture leading to each nest. The hen is remarkably prudish, and likes to be quiet and retired while laying. Dark nests often prevent the eating of eggs, of which bad practice hens confined are not unfrequently guilty. The roosts must *not be above two feet* from the ground. The common plan (placing the roosts in the highest part of the house) is *decidedly wrong*. The nearer the roof they are, the fouler is the air the fowls breathe, and besides, ventilation becomes impossible. Moreover, hens are often seriously hurt, and their feet destroyed, by flying from high roosts in small and contracted houses. Roosts should be flat, about six inches broad. Round roosts, so generally used, are apt to cause the breast-bone of *all* fowls, and especially *heavy ones*, to be crooked. For Brahmas and Cochins a *broad shelf* should be provided—otherwise they will take refuge in the nests, which ought not to be allowed.

Adjoining the hen-house there must be a covered shed, free from draughts, into which fowls may go in wet weather. The keeping of hens *dry* is so essential that this shed should have a wire front, with a door which can be closed at will.

Cobbett says:—" Hens should be kept in a warm place, and not let out even in the day time, in wet weather, for one good sound wetting will keep them back for a fortnight. The dry cold, even in the severest frost, if dry, is less injurious than even a little wet in winter time. If the feathers get wet in our climate in winter, or in short days, they do not get dry for a long time; and this it is that spoils and kills many of our fowls." From long experience, I have found that Cobbett is perfectly right in his remarks, and that in winter the protecting of the hens from rain, has a material influence in filling the egg-basket.

Attached to each hen-house *must be a yard*, as large as possible, and if practicable, one part should always be retained in grass; or there should be a field, with an opening from the several yards, into which each breed may be allowed to go for a few hours each day. Fowls are very tractable and docile, and after a day or two will recognise their own yards. They seem to know *too*, that they must make the best use of the short liberty given, for they devote all their energies to picking up grass, insects, worms, etc., and perhaps do more in that brief hour of freedom than they would do, if allowed complete liberty the whole day.

In a farm-yard where there is plenty of space, and more especially where only one breed is kept, fowls may have complete liberty. "A farm-yard," it has been well said, "is the paradise of poultry." Still, even on a farm, much might be done to make a better use of its advantages. Besides his permanent hen-houses, the farmer should have portable houses, which he could place in any part of his farm as occasion required. In autumn, and during winter, these houses might stand in stubble land, or even during the whole year on these fields which are laid down in grass; ample range would thus be given to the poultry, while at the same time the ground would be improved by the manuring it would receive. The kind of house I would recommend, would be *in size* according to the number of fowls kept in each flock. As it would be necessary to remove it from field to field, it should be placed on wheels about two to three feet in diameter, with flaps of wood on hinges at the back and on each side, which could be easily lowered. This would form an excellent shed for the hens in wet weather. The house itself should be provided with nests and roosts, and be reached from the ground by a ladder formed of a plank, with bars of wood nailed across, at distances varying from six to ten inches. If reared from chickenhood in this way, poultry would lay the foundation of most excellent constitutions, which would stand them well in our variable climate.

The hen-house should have a south-eastern, but in no case a northern exposure, and the ground on which it stands should be slightly raised, having a declivity immediately in front. This will help to keep the ground from becoming a puddle in wet weather.

Immediately after the fowls leave the hen-house in the

morning, the door and windows or ventilators must be opened wide, so as to admit as much air as possible. The floor should then be swept *perfectly clean* with a birch broom, or if this does not clean it thoroughly, it should be scraped. A little dry earth sprinkled on the floor will serve the double purpose of deodorising the manure, and of preserving it in good condition for use in spring. Once a fortnight some dry lime should also be scattered on the floor. This will give a healthy smell to the whole place. Twice a year the whole building should be thoroughly and completely white-washed. Every crevice and corner should be searched by the brush, so that no dirt could by any possibility remain. If these precautions are attended to, insects need not be feared, but if (though, after what I have said, it will be the poultry-keeper's own fault,) these pests of the hen-house find an existence in it, an immediate effort must be made to get rid of them. Chloride of lime, paraffine, or a solution of carbolic acid, will be found effectual in expelling them, if judiciously used.

The roosts should also be carefully washed, once a fortnight, with clean water in which lime has been mixed, and a little sulphur scattered over them. The nests must not be forgotten. The hay and straw which are intended to make them more comfortable, must be very frequently renewed, and the fresh supply sprinkled with a small quantity of sulphur. Oat straw makes decidedly the best material for lining the nests, but whether this or any other is used, it should be broken up into fine parts, so as to prevent the fowls catching it in their feet while leaving the nests, and to do the same. If poultry-keeping, as a separate branch of industry, is found in France to be profitable, it should most unquestionably be found so here. We have equal facilities, and, being on the spot, we can command an immediate market, and forestall the necessity for importation at all. Besides, even if we could not at once check importation, we need not fear it. The demand is so very much in excess of the supply, that, even were the supply doubled, we would, with an increasing population, be still unable to meet it. There is not the slightest doubt that any man of ordinary intelligence, if possessed of a small plot of ground, would make a respectable livelihood by cultivating more closely and carefully this branch of industry. One great advantage it offers to everyone is, that almost no capital is required, and that what little capital is employed, is being continually turned over, each time realizing a high percentage.

PART I.

perhaps taking the whole contents of straw and eggs with them. A china egg should always remain in the nest.

The feeding dishes *must be well washed daily;* and the water vessels *must twice a day* be thoroughly rinsed and cleansed, and renewed with fresh water. A short heather brush will do the work most effectually.

The yard must also be subject to the reign of cleanliness. From the constant droppings of the fowls, it is very apt to become foul, and tainted ground is the forerunner and cause of disease. It must be raked daily with a very fine tooth rake. Once a month the whole yard must be dug to the depth of a foot, and the earth turned over; and once a year, six or twelve inches of the surface must be renewed, and fresh earth put in its place. What is taken away will be found most valuable for manure. It should be kept in a dry covered place, free from the influences of our changeable climate.

In one corner of the covered shed *must be placed* what is called a dust-bath, composed of fine dry ashes, or sand and lime rubbish. Fowls, even under the best treatment, are subject to much annoyance from the presence of vermin, and from this they can only free themselves by rolling in the dust bath. A shallow box is quite sufficient to serve the purpose, and its contents should be mixed with a few pounds of black sulphur.

The houses being good and perfect, the next important point is cleanliness, and it is attention to this, *in all its details*, which more than anything else will contribute to the prosperity of the poultry-yard.

CHAPTER II.

THE BREEDS.

ONCE the houses are ready for occupation, the next thing to be done is to find the occupants. It is to be hoped that the intending poultry-keeper has been keeping this in mind, and that he is quite ready with the particular breed or breeds he wishes to adopt. As my treatise is intended for the use of those who are not as yet supposed to know anything about the different breeds, I shall briefly enumerate and describe the most important of these, stating at the same time the kinds I would recommend as being most likely to return large profits. Of course the first thing that must be decided is whether eggs or chickens is most desired. There are some breeds whose special characteristic is laying, and who would be perfectly useless as either sitters or mothers, while there are others whose chief recommendation is that they make most excellent sitters and irreproachable mothers. Perhaps the best way is to divide the breeds into the two classes of *sitters* and *non-sitters*.

For sitters, we have Dorkings, Game, Brahmas, and Cochins. For non-sitters, we have Hamburghs, Spanish, Minorcas, Andalusians, Houdans, Polands, Crêve-coeurs, La Flêche, and Leghorns.

The Dorking is a plump, compact, and well-shaped fowl, and is possessed of a fifth toe. It looks exceedingly pretty when in full feather. It eats largely, and has of all fowls the greatest tendency to fatten. It is both an excellent sitter and mother. It is decidedly the most perfect table bird we have, combining quantity with quality. Of all dead poultry sent to the London market it carries off the palm. The largest price is given for it. Sometimes as much as 8s. and 10s. during the great race week. Unfortunately it is a very delicate bird, and totally unsuited to confinement. The chickens are exceedingly difficult to rear; and if the soil is in any way damp or moist, few of them will come to maturity. Dorkings are most indifferent layers, especially in winter, and after two years they almost cease to lay altogether. They age very rapidly, and are subject to an inflammation of the foot, known as "bumble-foot." There are several varieties, named respectively white, speckled, grey, and silver grey.

Game fowls are very hardy, and the most courageous of all poultry. They are small eaters, and though more fitted for a large run, can easily be kept in confinement. As sitters, and afterwards as mothers, they are not to be surpassed; and no hen can protect her brood of chickens from danger more cleverly. The chickens are hardy and very easily reared. They seem, on little food, to form as robust constitutions as any other variety. Though the eggs the game hens lay are small, they are very delicious, and are very abundantly supplied. As a table bird, they are unequalled for richness and delicacy of flavour. The one great drawback to game fowls is their pugnacious disposition. The names of the several varieties are duckwings, piles, black, and brown-breasted red.

Brahmas are a particularly useful and beautiful breed, large, very hardy, and suited to confinement. The chickens are also easily reared. In winter, with slight protection, in the midst of frost and cold, it is possible to hatch and rear them successfully. The hens are most faithful sitters and mothers. They lay, *especially in winter*, a large egg, slightly tinted. Considering their broody propensities, they may be called excellent layers. They are very good table birds up to a certain age, being full, large, and meaty in the breast. After six months their flesh becomes inferior to the dorking. Light and dark Brahmas constitute the two varieties.

The Cochins are also most attractive birds, very suited to confinement, and, like the Brahmas, easily reared when young, and particularly hardy when full grown. Only while young, however, do they form a good table bird, as they increase in coarseness with age. They lay well. Their great recommendation is their constant desire to sit, sometimes as often as five and six times in the year, and never during each successive hatch, do they appear weary of their maternal duties. The varieties are known as white, buff, and partridge.

So much for the sitting breeds. Of the whole, I have no hesitation in giving the preference to the Brahmas.

Let us now consider the breeds classified under the head of non-sitters.

First and foremost we must rank the Hamburghs. They are the most abundant and prolific layers of any known breed, and at the same time are very small eaters. They are particularly hardy, and the chickens come early to maturity. Their eggs are not large. As table birds, though plump and delicately flavoured, they are of small size. A more beautiful breed than any one of the five varieties it is impossible to find, and on a lawn or at lodge gates they are most attractive. They will bear confinement well, but where they have perfect freedom their plumage will attain a beauty which is not surpassed, if even equalled by the pheasant. The names of the several varieties are the *silver-pencilled*, the *silver-spangled*, the *golden-pencilled*, the *golden-spangled*, and *the black-hamburgh*. Of the first four the silver-spangled lay the largest eggs. But I prefer and strongly recommend the black-hamburgh. It is now a distinct breed, though originally produced by crossing the golden-spangled with the Spanish. The result has been to combine the advantage of an enormous laying power with a large egg and a good sized fowl. It is besides, when in full feather, as beautiful a fowl as any one could wish to see.

The Spanish are a well-known and justly esteemed breed. They possess the special advantage of laying larger eggs than any other variety of fowl, the eggs weighing generally about three ounces, while not unfrequently they weigh four and a-half. The result is, that the eggs are always sure to sell first, and at the highest rate in an open market. They are moderate eaters. As table birds, their black legs are a great drawback, and are not compensated for, in the opinion of many, by the delicacy and flavour and beautiful whiteness of their flesh (all of which they possess). They are hardy when full-grown, but as chickens are difficult to rear. They are admirably suited to confinement, and from their dark plumage always look well, even in the darkest backyard of a manufacturing town. They have also what adds much to their beauty, a pure white face.

The two principal varieties of the breed are the Minorcas and the Andalusians. The Minorcas resemble the black Spanish in every way, but are a larger and more compactly formed breed. They hold the first place as layers of all the Spanish breeds, and their chickens are hardy. The Andalusians are the hardiest of all. As chickens they are strong, feather very quickly, and come easily to maturity. When full grown, they remain apparently indifferent to cold and frost. They lay very well in winter, which cannot certainly be said for their more aristocratic white faced relations, and are also much superior to them as table birds. To the names of the varieties already given, we may add, the white Spanish, the Anconas, and the Columbians, but as they partake more or less of the characteristics of the others we need not describe them.

The Polands are a remarkably handsome breed, being distinguished by a crest, or top knot of feathers, and taking high rank as a purely fancy fowl. They thrive most under petting, and manifest what few fowls do, attachment to their feeders. For market purposes they are not of sufficient size, though they fatten easily, and their flesh is white and tender. In spring and summer they are very good layers of large white eggs. Unfortunately, they are delicate, and do not thrive on a damp soil, suffering particularly in wet weather. When their top knots get saturated with the rain, it is apt to bring on cold and roup. They are timid, and as their crest obscures their vision, should never be seized suddenly without some indication of being approached. The *gold* and *silver-spangled*, and the *black with white top knots*, constitute the several varieties.

The Houdans are one of the varieties of fowl introduced from France. I consider them *the most valuable addition* to the poultry yard, and believe that in the course of time they will, when their great and good qualities are better understood, take high, perhaps the highest, rank as domestic fowls. They combine the very necessary qualifications of being splendid layers of large-sized eggs and very precocious in fattening, as well as being particularly hardy and easily reared. The chickens feather quickly, and, if the Brahmas and Cochins are excepted, are not surpassed for strength of constitution and general stamina by any other breed. It is the exception ever to find a Houdan sick, and, no matter how small or confined the space in which they are cooped, they will thrive well, and give in eggs and flesh an excellent return for good feeding. They are handsome, of black and white spangled plumage, and have a large crest surmounting the head. They have, like the Dorking, a fifth toe.

The Crève Coeur is another of the French breeds. They are the breed of all breeds most suited for confinement. Even when unlimited range is provided for them, they seldom wander from their roosting house. They are beautiful birds, with rich glossy black plumage and large black crests. Their combs are curiously divided into two well-developed horns. Their *special merit* consists in the rapidity with which they fatten. Chickens may when ten or twelve weeks old be quite fit for table. The quantity and quality of their flesh are not equalled by any other fowl at a similar age. They lay large eggs—equal in size to the Spanish—and continue laying during the autumn when most birds are deep in moult. The home-bred birds are stronger and hardier than those imported from France.

The La Flèche is considered one of the finest fowls in France, but, I think, is not suited for our variable *climate*. The chickens are delicate and difficult to rear. The eggs of the La Flèche are large. The flesh is delicate, well-flavoured, and white. Like the Spanish, they have black legs, and are not unlike in general appearance. The plumage is dark black with green reflections. They have long legs, and are much afflicted, particularly the cocks, with leg weakness and disease of the knee joints.

Before passing from the subject of the breeds, I should like to name the Leghorns, a breed lately introduced from America, and which is found to be hardy, plump, and a prolific layer of large white eggs.

In giving a description of each of these breeds, I have confined myself to those *useful qualities*, which should commend themselves to the consideration of poultry-keepers and those for whom my treatise is specially intended.

CHAPTER III.

FEEDING OF POULTRY.

THIS chapter treats of one of the most important subjects connected with the management of poultry. Generally speaking, poultry are given, at *one meal*, what without scrimping would do for at least six.

"The tendency of over-feeding," says Baily, "is to make fowls squat about under sheds and cart-houses, and instead of spreading over a meadow or stubble in little active parties, searching hedges and banks, and basking on their sides in the dust, with opened feathers, and one wing raised to get all the glorious sun's heat that they can, they stand about, a listless pampered troop. To lay *much better*, to breed better chickens, and to last longer are the results of diminished, not increased, expense; and all that is required is a little personal superintendence at first, till the new system is understood and appreciated. In most yards the birds are *over-fed*, and there is waste in nearly all."

While it is most essential to be careful *not* to over-feed, the other extreme of under-feeding must be avoided. I believe that one reason farmers' poultry "do not pay," as they themselves declare, is that they are starved—allowed to feed *when* and *how* they like—and are the only animals about the steading which are neglected and not daily supplied with regular meals.

As a guide to the quantity required for each hen, let me here say that only *as much as*, and no more than is necessary to maintain the birds in health, and to keep them active for laying, should be given. The kinds of food must likewise be judiciously selected. To give grain two or three times a day, without soft or green food, is wrong. To feed entirely on soft food is equally wrong. A wise discretion must also be exercised as to the materials which form the soft food. Some meals have a very fattening tendency; others again, while imparting nourishment to the system, are favourable to the formation and production of eggs.

Great attention must be paid to the hours at which the fowls are fed, and to the nature of the food given at each diet of feeding. If the fowls have complete liberty, not only will *less food* do, but *less soft food* and more grain may be allowed. If they are constantly penned up, a larger supply of soft food must be given. At all times, the feeding must be attended to with the *utmost punctuality and strictest regularity*. One day there must *not* be starvation—the next over-feeding; but every day the food must be given regularly, in equal proportions, though not of the same variety.

FEEDING OF POULTRY.

If fowls have perfect liberty, it is not necessary to feed them more than twice; and though it is better to give them three diets when penned up, yet the only one that should be a full feed must be the first. The other two may be reduced to half a feed each.

The first meal must consist of soft food. This may be composed of barley-meal, oat-meal, buckwheat-meal, Indian-meal, bran, or sharps, along with potatoes, potato peelings, turnips, mangold wurtzel, and all other scraps of fat, suet, meat, which may be over from domestic use. A very good food may be obtained by having heavy oats or barley crushed whole and mixed with the husks.

A variety of feeding is the very life of poultry, and by a judicious selection of the above this may always be procured. Some of the foods recommended can only be used in conjunction with others. For instance, either bran or sharps would be too porridgy if prepared without any other meal; Indian-meal would be detrimental to the production of eggs, as its tendency is to fatten; and potatoes alone are not to be recommended as a regular diet. When, however, potatoes are mixed with bran or meal, or when the Indian-meal is conjoined with one of the other meals, a most excellent diet is obtained, which cannot be surpassed, and which will be found most conducive to health and laying.

The poultry-keeper should try to make himself acquainted with the relative qualities and ingredients of the different kinds of food. Each season requires a change in the feeding; thus, in warm weather, when the production of eggs is greatest, that food should be given which contains most nitrogenous or flesh-forming material, while in cold weather that food which contains the most carbonaceous or warmth-giving and fattening properties should be supplied.

Raspings of bread, which can easily be obtained from any baker, are a most excellent substitute for sharps, and particularly when mixed with fat or boiled with liver. Cummins, which may be had cheap at a brewery, are also a most valuable addition to poultry food.

But perhaps *one of the best*, if not the *very best*, food is malt. I believe, expensive as it is, that it will repay the poultry-keeper by inducing an early and constant production of eggs. It must be used with discretion; a small quantity (one handful to every three fowls) mixed with the ordinary food is all that is required. It may be used either whole or ground.

It is necessary in winter, in the case of fowls penned or not, and at *every* season in the case of fowls in confinement, to give them a little animal food daily, as a compensation for worms, insects, grubs, etc., of which they are very fond, and without which they will not be kept in good condition, or yield a regular supply of eggs. Bullock's liver or sheep's pluck can be had at a small cost. They should be thoroughly boiled, and then chopped up very fine, before being used — not more than half-a-finger-length being given to each fowl daily. The water in which the meat is boiled should be utilized in mixing the food. All kinds of bones should first be boiled in it, so as to increase the strength of the broth. The bones should afterwards be crushed very fine and given to the hens. It is not advisable to give them *raw* animal food at any time, as this generally induces them to begin the pernicious practice of eating each other's feathers, which would soon disfigure and destroy their beauty.

In addition to this, green food must be daily given, especially to all fowls penned up—grass, cabbages, lettuces, in short, any kind of vegetable should be chopped up and mixed with their soft food—due care being always taken that an equal quantity is given, so as to prevent any disorder arising from a sudden increase of it. Besides giving green food in this way, it is a most excellent plan to hang up a cabbage, a turnip, or mangold wurtzel in the shed, about nine inches from the ground. The eager rush that will be made to it the moment it is put in its place, is quite sufficient proof of the relish which the fowls have for it; and the whole day some one of them will be found constantly employed in picking at it. Attention to this last simple suggestion will be eminently conducive to the health of fowls in confinement, as it gives them some kind of occupation and exercise.

We must now refer to grain. There are many kinds which may be employed, and which are easily obtained, such as Indian corn, wheat, oats, barley, and buck-wheat. The first named possesses a great tendency to fatten. In its use great care must therefore be exercised, and perhaps it is better to give it every alternate day at the half feed in the middle of the day, or when the weather is wet and cold. Even then, the smallest possible quantity must only be given to each fowl.

Wheat is a most excellent and valuable addition to poultry food. The preferable way to use it is after boiling it. The only objection to it is its high price. Oats are, when heavy and good, an excellent substitute for other grains. Barley is, with the exception of buck-wheat, unsurpassed in its value for poultry, and may constantly be given either whole or steeped. In the latter way, fowls are immensely fond of it. Sometimes, too, it is a good plan to boil it.

Of buck-wheat, I cannot speak too strongly. It is unquestionably *the food* for poultry. It seems to stimulate the laying powers of the hen, and to keep her, notwithstanding the exhaustion which must attend a great production of eggs, in good condition. It has not as yet been very largely grown in Britain, though I think its valuable properties are already recommending it to not a few farmers. It is in colour dark, and of a small triangular shape. When ground, it makes a most excellent meal, which may be substituted for other meals in the mixing of the soft food. To the above grains I must add hemp-seed. It is *too expensive* to give as an ordinary food, and even if it were less so, I do not consider it would be beneficial for fowls. During the moulting season, however, it is invaluable, and should be given every second or third day. At this particular period—the most critical in the life of the fowl—it is absolutely necessary to supplement its own efforts by high and stimulating feeding.

Water must *never* be forgotten. It is one of the most essential requirements of the poultry yard, and should always be abundantly supplied. It must be renewed in winter once, and in summer twice a day. At each renewal the fountains must be washed and cleansed, and placed in a cool shady spot when the sun is strong, and under cover when the weather is wet. Nothing is more injurious to fowls than rain water. In wet or cold weather, in winter and during moult, a solution of sulphate of iron and sulphuric acid (one pound of the former, and one ounce of the latter, dissolved in about a gallon and a-half of water,) should be mixed with the drinking water in the proportion of one teaspoonful to each pint, just sufficient to give it a kind of mineral taste. I have used this mixture largely, and am never without a jar of it. I give it once a week to *all* my fowls, and find they are the better of it. The recipe was originally prepared, and afterwards published by that eminent poultry-breeder and fancier, Mr. John Douglas, in "*The Field*," and has been called after him, "*Douglas Mixture.*" It is not expensive, and should be used—once a fortnight in summer, once a week in winter, and oftener during moult.

A little lime-water may be added to the water once a week. It will help to form the shell of the egg. It is easily prepared by having a barrel *full* of water, in which may be placed a lump of limestone.

Whatever kind of vessel for the water is selected, it should most certainly be one that can very easily be cleaned in every part. The best vessels are either flower pots, reversed in their own saucers, or flower pots alone, with a cork inserted in the under hole. They are cheap, and what is most essential, they are easily and quickly scrubbed.

CHAPTER IV.

HATCHING AND REARING OF CHICKENS.

Two of the most interesting duties connected with the poultry-yard, are the hatching and rearing of chickens, and yet there are none more difficult. Failure in both is sometimes due to the weather, or to some extrinsic circumstance over which, even the most careful, can have no control; but it is, generally speaking, due to the neglect of those *necessary conditions*, without which success is impossible.

Lately, when so much attention has been paid to the management of poultry, an attempt has been made to supersede the sitting hen by the introduction of incubators, but I do not think their adoption is ever likely to become general. The keeping up of an even temperature, while giving sufficient moisture to the eggs, makes their management a very difficult thing. Even after the chickens are hatched, difficulties increase, for it is almost impossible to rear them successfully. What is called artificial mothers, have been designed to take the place of the natural mother, but these have always been attended with indifferent success.

It is a well-known fact, that both in Egypt and China chickens are hatched most successfully in very large numbers; but the climate of both these countries makes the hatching an easier and much more possible thing. Moreover, in the case of the Egyptians, hatching has been a regularly recognised profession—the manner in which it is conducted being kept a profound secret by the particular families who engage in it. All that is positively known is, that the hatching takes place in ovens under ground. One thing is very certain, that all successful hatching by artificial means, is entirely dependent on the close imitation of nature; and I believe the success of the Egyptians is due very much to this, after long and patient and laborious study.

I need not enter further into the description or use of incubators, as those for whom my treatise is intended are not in the least likely to use them. When used, I have always found that they are soon thrown aside.

A good heavily feathered hen is the very best possible incubator, and if Brahmas or Cochins are kept, there will be no difficulty in having some of them broody at all seasons, and more particularly early in spring, when hatching should be pushed forward.

Barn-door fowls make most excellent mothers, so that there can never be any difficulty in finding hens ready to hatch eggs. In choosing a hen, it is always important to select one that is in good feather, and of quiet and gentle disposition. Young hens or pullets are generally at first restless, and do not always sit faithfully throughout, though with kind and gentle treatment these faults may be overcome.

It is very essential to be careful in the selection of eggs. None should be set where there is the slightest doubt of fertility. Some people assert that if the air vessel at the large end of the egg be visible, it is fertile; while others declare that it is possible to distinguish the sex of the future chick by the position of the air bubble—that if it be in the centre of the blunt end of the egg, a cockerel, and if it be at the side, a pullet will be hatched. It is, I think, hardly necessary to add that these theories are absurd. Eggs having any oddness in their shape, or thinness in their shell, must be avoided. They will not hatch.

I condemn the practice of allowing hens to sit during the important period of incubation in the *laying houses*. The presence of other hens always makes them restless and unquiet; and very often this, more than any other reason, causes the clucking fever to disappear. Large sitting-houses are objectionable, inasmuch as while feeding, hens being very pugnacious, are apt to fight, and in the fray possibly eggs may be broken, or the hens become *too excited* to resume their places. A sitting-house should be subdivided, so that not more than four or six hens can be fed at one time.

Decidedly the best plan is to have for each hen a wooden pen, measuring six feet long and two broad. The cost is not great, and it has the advantage of being useful after hatching for the rearing of the chickens. The house should be *two feet* square, while the court should measure *two feet* by *four feet*. The court, as well as the house, must be covered in, and, in view of the rearing of the chickens, it should be enclosed on three sides—one side only having a wire front. The nest should be placed in the house; and food, water, and a dust bath in the court, where *once a day* the hen should be allowed to go.

Another great advantage of these pens is, that being without any bottom, the nests can be constructed on the ground, which always assists incubation. When the nest is made in a box, a layer of fresh turf must be placed first in it—care being taken that it is large enough to fill each corner. A nest must be made by scooping out the centre, and lining it with soft oaten straw, which is preferable to either hay or coarse barley straw. Fresh cut grass, moss, or heather, are all very excellent for lining nests. If the small hatching houses or boxes I speak of are adopted, all that is necessary is to shape the nest out of the solid ground before lining it. It is also very necessary to be particular as to the choice of the site of the sitting box.

In January, February, or March, it is better to have it placed in a warm, dry, sheltered spot, for during these months eggs are very apt to get chilled, and a chill will endanger a whole setting. Later on, a cool damp situation should be found, as in warm weather moisture is essential to successful hatching. Keeping in view these two re-

HATCHING AND REARING OF CHICKENS. 49

quisites, it will be easy to find suitable situations for the hens according to the time of the year.

The season of the year must also regulate the number of eggs placed under a hen. In January and February no more *than seven* should be set under each hen; in March, *nine* may be allowed; in April, *eleven*; in later months, *thirteen or more*. At no time should more eggs be put under a hen than she can possibly cover; for as she constantly changes the position of the eggs, each one in turn is thrust outside, and is most likely to get chilled, and thus the whole setting may be lost. Even though a hen may be able to cover all the eggs, it should not be forgotten that it is impossible for her to give sufficient covering and protection to the chickens when hatched, and this is absolutely necessary if they are to be successfully reared during the early months of the year.

The hen sits for twenty-one days before the chickens are hatched. Though this is the recognised time, yet not unfrequently they may hatch six or eight hours earlier, or perhaps twenty-four or forty-eight hours later.

An excellent plan is to set two, four, or six hens at the same time. When clucking hens are scarce hens may be made to bring out a second brood of chickens. The best way to accomplish this is to divide all the chickens hatched by three hens between two, and to leave the third free to sit a second time. If fed well, and carefully attended to, there is no hardship to the hen in this arrangement.

It sometimes happens that the chick takes a long time to free itself from its shell. In any such case, if assistance be rendered, it should always be of the scantiest kind, for the slightest flowing of blood is fatal to the chick. The best thing to do, is to insert the point of a pair of scissors into the hole made by the chick in its effort to escape, and to cut up the egg towards the large end. Place the chick immediately under the hen, and though there is *no guarantee* that its life is saved, yet all has been done that could be done to preserve it.

I have, however, invariably found that chickens *hatch best* when the eggs are *touched least*.

The first food of the chicken should be an egg hard boiled, and chopped up, with a few crumbs, and mixed with a little warm milk. This, with the addition of a little animal food rather underdone and chopped very fine, or even a few worms, should constitute the diet for the first week. If the chickens are hatched early, it will be necessary to feed by lamp or candle light at 11 o'clock at night. The best way to accomplish this is at first to lift the hen gently but firmly from the nest, taking care that none of the chickens are lifted along with her, and then dipping the bill of each chicken in the food. They are very apt to learn, and soon will be as eager to get the food as anyone can be to bring it.

By the end of the week the egg may be dropped, and recourse had to ordinary meals, such as have been recommended for other fowls. The best of all is malt. I strongly recommend in addition Spratt's or Chamberlain's prepared or patented meal, and Thorley's food for cattle. The use of these meals increase at the time the cost of rearing the chickens, but I believe it ultimately will be found more economical. Dry bones crushed into a fine powder or meal should be daily given mixed with the other food, and the result will be the rapid feathering of the cockerels, and the early laying of the pullets. Chickens should always be satisfied at each feeding, but the moment they cease to run after the food thrown to them, or seem careless about it, it should be removed. Whole grain, and especially hemp-seed and buck-wheat, should be given after the first week as the *last meal at night*. Grass should also be cut fine and mixed with their food, and if this cannot be conveniently got, any kind of vegetables, but more especially *lettuces*.

Mashed potatoes, turnips, vegetables left from the dinner table, scraps of meat or pudding, mixed up in the water in which the dinner dishes are washed, form an excellent and nutritious food. Rice is not to be surpassed as a food for chickens. It should be allowed to simmer at the fire all night. It may often be purchased very cheaply when slightly damaged, and then is perfectly good for use. Warm milk given during the first month early in the morning, will also be found beneficial. The water must be provided in shallow dishes. During the whole time the hen is brooding, she should not be stinted in her allowance of food, as it is necessary to get her as soon as possible into condition. It must never be forgotten that the moment her maternal duties are ended, she is to be sent back to the poultry-yard to lay eggs, and that her usefulness and profitableness depends, for a time at least, on her condition when returned to the yard.

We have already spoken of the coops for the chickens, and recommended that the yards should be covered in and enclosed on three sides with a frontage of wire. In the case of early broods it would be of immense advantage to have the half of the wire front incased in glass, as especially during January and February, the frost and cold are sometimes very severe, and the wind keen and biting. The construction of a temporary glass front, which may be removed at will, during the late spring or early summer, when the fresh air is most desirable and beneficial, is both simple and cheap. Almost every wright or house-carpenter has old windows lying beside him, (removed from houses that have been demolished), which he will sell or fit up at a very trivial cost. *In no case* should the coops have a bottom. Wooden floors are particularly injurious to chickens, and, unless they are seized with cramp, should never be used. When seized, a temporary removal to a boarded or wooden floor will be found beneficial.

Though chickens have often been reared successfully without any sun at all, it is of great benefit to them. Hence a window in their coop should, if possible, open to the south or south east, through which the early sun may enter.

During the whole period of chickenhood the greatest possible attention must be paid to cleanliness and regularity of feeding. Chickens must not be allowed to roost for at least three months after hatching, and even then all small roosts must be avoided. A broad shelf littered with straw, about six inches or a foot from the ground, or the flat roost previously recommended for hens, is decidedly the best for them. It is the nature of all fowls to rise early, but it may be very judiciously checked for at least four months of chickenhood, by keeping the roosting house well darkened, till about seven o'clock in the morning. After they are a month old, chickens do not require such early feeding as is necessary at a tender age. And nothing is then better for them than long hours of rest and sleep.

PART II.

CHAPTER I.

GENERAL SUMMARY.

THE first remark I must make, and *I consider it a very important one*, is, that whether few or many hens are kept, the whole affair must be considered in a *business* point of view. Food must always be bought *wholesale*, and given to the fowls *without any waste;* and the necessity for attention to both of these injunctions will be all the more readily seen if there be, as *undoubtedly there ought to be, personal supervision* on the part of the poultry-keeper. A few minutes daily will overtake all that has to be attended to, and it is, I think—on the principle that it is better to do *well* whatever is undertaken—not too much to ask or expect.

I do not know that I can give a better advice to the amateur poultry-keeper than this one, that it is better, before beginning poultry-keeping on a large scale, to gain experience by keeping a few common barn-door fowls. He will very soon learn all that is necessary to keep them in good condition and health, to have a constant supply of eggs, and a good bird always ready for table. Once he has obtained all necessary information, he must *at once* cast aside the common fowls and begin with *pure breeds*. Some have ventured to assert that barn-door fowls may be made as profitable as the best and purest breeds; and, no doubt, under more careful management, their laying powers may be very greatly increased, their size of body enlarged, and their flesh rendered more delicate. But it is impossible, whatever the management, to attain the same great results with them as with the pure breeds. From long breeding in and in, with no fresh blood introduced, the barn-door fowl has deteriorated into one of the puniest and most insignificant of creatures; and possesses, with half the capacity to fatten and to lay, an enormous appetite, and eats, I have found from experience, a great deal more than its aristocratic relations.

In the selection of breeds, it is necessary to remember that each has its distinctive merits and qualifications. *Some hens do not sit—others do*, etc. In an ordinary yard, as in the case of cotters or families, where an abundance of large fresh eggs is desired, I would recommend either the Houdan, Crêve-coeur, Andalusian, black Hamburgh, or Leghorn. Of course every one knows what beautiful eggs the black Spanish lay, and perhaps wonder why *they* are not suggested. My objection to them is their extreme delicacy, rendering them always most indifferent layers in winter, and, besides, possessing such small bodies that, though well flavoured, they cannot be considered good table fowls. The Andalusians are a variety of the Spanish, and are

GENERAL SUMMARY. 59

infinitely stronger and hardier. They lay an equally large egg, and form a better sized bird for the table. The Houdan and Crêve-coeur are not to be surpassed as table birds. It almost seems fabulous the size they attain on less food than common-fowls, and I have had hens often weighing eight pounds, and cocks ten. They are quite as good for table use as the much-recommended Dorking, and they possess this further advantage, that neither as chickens nor as adults are they in the least delicate. At the age of twelve weeks—so early do they fatten, and so quickly do they attain maturity—they are fit for table. They are also splendid layers. The Leghorn has not the same size of body, and cannot therefore be reckoned an equally good table bird, but it is a first-class layer. The black Hamburgh stands unrivalled as a producer of good sized eggs. All the Hamburgh breeds lay splendidly—lay more eggs in a year than any breeds known. They are also remarkably small eaters. If fowls are wanted as sitters, I would recommend either the Brahma or Cochin. The former lays larger eggs, and is more tender as a tablefowl. The latter will sit as often as six times in a year. A cross between the Dorking and the Brahma, or the Brahma and the Houdan, makes an excellent table fowl. Of all the breeds, none are more suited to confinement than Crêvecoeurs and Houdans as non-sitters, or Brahmas and Cochins as sitters.

If the accommodation can at all afford it, I most strongly urge the keeping of two breeds, the one for laying and the other for hatching.

The stocking of the yard or yards is very easily done, if the amateur has plenty of money, as he can then readily purchase approved breeds from well-known breeders. But, as each fowl will in this way cost from fifteen to thirty shillings, I recommend that he should either buy a few stock birds—say four hens and a cock—of each of the two or more breeds he intends to keep, and hatch *all the eggs they lay;* or he should, at the breeding season, purchase setting eggs from some *well-known breeder*. I lay stress upon *well-known* breeders, and I *distinctly* and *decidedly* advise all to deal with them. They certainly charge a high price, ten, fifteen, and twenty shillings a setting, but, if one pays well, he has the satisfaction of knowing he is honourably dealt with. Many of those who advertise eggs and fowls are, as I know from sad experience, not to be trusted, though there are honourable exceptions. The difficulty always is, to know who are the honest and who are the dishonest advertisers, for often the advertisements are most misguiding—the most pretending being, generally speaking, the least to be depended on. Eminent breeders, on the other hand, are too conservative of their honour and name to allow anything to leave their yards which could in the least be considered doubtful.

If, instead of purchasing eggs at the breeding season, it is determined to start with a few stock birds, from which all the eggs are to be set, it is very essential to have the yards made up for breeding, if possible, in *November*, and at latest in *December*. I have found this hasten laying during those months when eggs are most scarce, and especially will eggs be plentiful for setting in the months of January and February. It is very important—I may with truth say *the most important thing* connected with poultry-keeping—that the proper time for hatching be remembered. Many people who keep fowls hatch at all seasons indiscriminately. The result is at once apparent, that, with the exception of the few chickens that by a lucky chance were hatched early in the year, all the rest turn out unprofitable. It is the *winter supply of eggs* that controls the profit of poultry-keeping; and, therefore, the first and *pre-eminent* question for the amateur to decide, is how this is to be obtained. As a rule, hens begin to lay when six months old. Chickens hatched in January, February, or March get over their moulting early, and have the advantage of the summer sun to hasten them to maturity, and will begin to lay at the season when eggs are scarce, and continue to do so during the dear months of winter, when a really fresh egg is a positive luxury. Moreover, if the eggs are sold in winter, they will realize a much larger price than at any other season of the year, and as this fact is one of the results of bad management *entirely*, those who choose to take the trouble may, with perfect ease, reap the profit and benefit of their care and forethought. Many people hatch in the latter end of June or perhaps July. The result of this is that the chickens, retarded in their growth by the cold and frost of winter, do not begin to lay till the March following; and the expense of keeping them for ten months without any return, and of only obtaining eggs when eggs are cheap, renders the whole affair most unprofitable.

Let it then be established *as a first principle,* that during January, February, March, and April *only*, are chickens to be hatched for the purpose of getting winter layers. In June no eggs should be set. It has been remarked by careful observers that eggs set during June do not always hatch, and that the chickens which are hatched are delicate, and are afflicted with a kind of skin disease which generally proves fatal.

In July and August it is well to set again, as doing so rests those hens which it is desired to keep as stock birds, at the period when they have most need of rest. It also hastens their moult. The chickens hatched during these months come in for the table at a time when a chicken is a luxury; and may be found at Christmas a not unacceptable alternative to the goose which is so expensive. I recommend either the Houdan or the Brahma, or a cross between the Brahma and the Dorking, as the most likely of all breeds to grow quickly, unimpeded by severe weather, or by cold and frost. Ducks are always hardy after the first fortnight, and are to be specially recommended for late hatching. They feather and grow with remarkable rapidity, and seem in no way to be affected by the wet wintry weather. Ducklings and green peas at Christmas are truly a delicacy. After April, all eggs should be set with the object of providing chickens for the table or for the market.

It is not generally known that the hens of those breeds which sit, may be made to sit at any time. When this is desired, one must be selected which has previously brought out chickens—full feathered, and of a quiet gentle disposition. A comfortable nest, with a sitting of addled eggs, in a dark room, must be provided for her, and when set on them, a wire cage, large enough to cover herself and her nest, must be placed over her. If she proves refractory, the plucking of a few feathers from her breast, will induce her to seek to allay the irritation caused, by *sitting more closely on the eggs*. She must be fed, regularly and abundantly, on most stimulating food, such as buckwheat, hempseed, meat, and bread saturated or mixed with beer. All her excrements must be removed each day. In the course of a few days she will be found perfectly ready to continue sitting, and fresh good eggs may be entrusted to her, while the use of the wire-covering will be found unnecessary. If either of the two sitting breeds I have recommended are kept, this resort to compulsion will not be required.

Hens may be prevented from sitting, if they should manifest this desire too frequently, by shutting them, in a small dark box for a few days in a corner of the ground, without any nest, and feeding them abundantly during the time of their confinement. Of course care must be taken that air is admitted. Some consider that this end is better secured by leaving the hen without food, though always giving it plenty of water; while not unfrequently the practice is adopted of dipping the poor creature in cold water. It is always well to know how the desire of incubation, if frequently recurring, may be prevented; but it should never be forgotten, that to debar the hen from sitting entirely, would be very injurious to her health, and destructive to her laying.

As soon as the chickens are hatched, every care must be taken of them for the first week or fortnight, and thereafter they must be pushed forward with the most stimulating food, so as to get quickly over their dangerous age, and to increase their general stamina and constitution.

At the age of three months, all cockerels should be in a fit state for table use, and must then be killed. Cockerels fatten very quickly during the first twelve weeks of their lives at a very little cost; for the next two months they generally fall off, while their appetites increase greatly, and very often they are not in such a good condition for use at the end of *five* as they were at the end of *three* months. It is a great mistake to fancy that the chickens which are commonly purchased in the market, or even reared at home, are a fair representation of the size or flavour or delicacy which may be obtained by careful breeding and management. Chickens at the end of three months may be got, I maintain, precisely *double* the size and *weight* of those which are generally at present obtained, and this too, I again, *without fear of contradiction, assert*, at a vastly smaller cost.

While it is thus necessary to push forward cockerels so as to have them fit for the table at the age of twelve weeks, it is equally important to hasten pullets to maturity by giving them stimulating food. Chickens hatched in January or February will be laying in August or September, sometimes in July. March and April pullets will lay in October and November, and from that time will, with proper management, continue to lay regularly till the following spring. After a short rest then, they will, if they be of the non-sitting breeds, commence again, and will continue to lay until the moulting season. The sitting breeds will be ready to hatch in January, and thus by the proper disposition of both, a never failing supply of eggs and chickens may be had during the whole year. During autumn and winter eggs are scarce, and consequently very dear, and seldom fresh. Very many of the eggs sent to market at this season have been collected when they were plentiful and cheap, and have been preserved in lime.

Moulting begins earlier with some hens than with others. Those birds which moult early should be most carefully attended to, as it is the critical stage of their life, and it is wise to give them a plentiful supply of stimulating food, so as to help them through it. Meat should be daily mixed in their food, while hempseed, buck-wheat, and beer will be found most beneficial in keeping up their strength. After a year's production of eggs, the constitution of the hen is exhausted, and it is no more than fair, as some acknowledgment of her great efforts, to give her any possible assistance which may help her over her moult. With due care and attention, there is nothing to prevent hens moulting in August, laying again early in November, and thus, at the very dearest season of the year, they are once again making a return for the generous treatment they received.

At the commencement of their second moult, *they must be killed without any scruple.* No hen should be kept *which is older than two years and a half,* and at this age they are still comparatively tender, but afterwards their flesh is tough and insipid, while there is a perceptible diminution in the number of eggs they lay. Hens which moult *late* in the *first* season should *not be kept for a second*, but with *unfailing regularity be killed at the first appearance of moult*. For they would not begin laying again till after Christmas, and it would be difficult to recover the cost of keeping them after that. The pullets will all be laying at the very latest in November, so that any birds which have only commenced to moult in October will not be missed. This regular system will be found *the secret of success*, and *in no way* should *be departed from if the largest profit is desired*. With the exception of fancy fowls, which are valuable for their points, etc., all hens cease to be profitable after they are two and a half years old, however well they may look.

If fowls are properly and systematically attended to, disease will always be a stranger in their yards. If, unfortunately, it does break out, the fault lies with the poultry-keeper. *To prevent disease, by care and attention to the poultry, is therefore the best advice that can be given.* Should disease, however, appear in the yard, the sick fowls must be killed on the first symptoms of its approach, as they are very apt to communicate infection to the others. I approve of killing the sick fowl rather than treating it. It is the most *economical plan*, and is the simplest and surest way of stamping out disease. If the poultry-keeper determines on attempting a cure, he must remove the diseased fowl to a warm and well-lighted house.

As I said before, it is easier to prevent disease than

to cure it, and as one means of keeping fowls in health, I recommend that once a week a little jalap, and once every four weeks some salts, should be mixed with their food —not more than *one table spoonful* to *every fifty* fowls. Castor oil is a most valuable remedy, and should be given to fowls on the first appearance of illness. The restorative, known as "*Douglas Mixture*," must be added to the water once a week.

During the months of April, May, and June, eggs are cheaper and more abundant than at any other season, and it is well to preserve as many as are not required. This may be done either by smearing them with butter or tallow, or by placing them in layers on end in a water-tight barrel, and filling it with a solution of lime and water and salt. The eggs must always be covered completely with the solution. They will, under this treatment, keep fresh for six months or even longer.

Such are the general principles on which poultry-keeping will succeed, and the person who is determined to make it so, *must master all these details*, and abide by them. Everything must be done with the utmost regularity and dispatch, and certainly no mere sentimentality must prevent the killing of the fowls when they have reached the limits of their profitable age.

CHAPTER II.

GENERAL TREATMENT.

IF the poultry-keeper desires to have his own breeding stock, the number of hens allowed to each cock must be limited. No absolute rule can be laid down, as a good deal will depend on the breed and the range provided. Some breeds are more fertile than others. No more than four (or six at the very utmost) hens should be allowed to the Spanish cock; while Houdan cocks, being the most active of all breeds, may with safety have nine or ten hens. Brahmas and Cochins should not have more than six hens; while Hamburghs may be permitted eight or nine.

This, in a general way, will give an idea of the number of hens which may be judiciously allowed to each cock. In confinement, cocks are never so active as when at perfect liberty. One rule which will always guide the poultry-keeper is the appearance of the hens. If they want feathers upon their backs, depend upon it the cock is over-treading them, and the natural remedy is to allow him

GENERAL TREATMENT.

more. A little observation will very readily enable anyone to see how matters stand. Young cocks are more vigorous than old ones, but in no case should they be kept after exceeding their fourth year. When fowls are not kept for breeding purposes, one cock may be allowed to run with twenty-five hens. It is quite true that cocks are not necessary to increase, or even keep up the laying powers of the hens, as it has been proved that hens lay as well without them as with them. But pullets come more rapidly to maturity, and look smarter if, after they have attained their fifth or sixth month, a cock is permitted to run with them. In an unlimited range the presence of a cock keeps the hens together.

Eggs for setting, placed in a basket full of sawdust or meal, with the *large end downwards*, will be kept in good condition for a month in cold weather, and for a fortnight in warm. The fresher they are the better will they hatch, and the hardier at first will be the chickens.

Sitting hens must be fed daily, and at a certain specified hour.

The *same person should always feed them*. Hens very soon get accustomed to their regular attendant, and the presence of another is apt to make them restless, if not obstreperous. The food should for the most part consist of grain. A little soft food is necessary, but for many obvious reasons it is better to feed most on grain. Plenty of fresh water and a dust bath must also be provided.

The hen should not be allowed to absent herself from her nest for more than twenty minutes, and early in the year, even this time may be reduced. The great danger of miscarriage in hatching is *not after* the tenth day, as many suppose, but before it. Then a chill will cause the failure of the whole hatch; and this chill is very generally caused by the hen remaining too long off the eggs. In a very cold day in January or February she should be prevented absenting herself from the nest for more than ten minutes. After the germ is formed there is not the same danger, though all throughout the twenty-one days great care is necessary. If, after feeding, the hen refuses of her own accord to return to her nest, she must be gently driven back, and if this fails, she should be quietly yet firmly lifted and placed on the nest. The door or flap should immediately afterwards be closed, and very soon will she settle down to her maternal duties.

During the whole period she is sitting, she should be kept in perfect darkness, with the exception of the short time she is allowed to feed and dust herself. Sometimes hens may refuse to leave their nests daily. The attendant in such a case must very carefully lift the hen out, for it is a cruel practice to allow her to sit for more than one day without food. This practice of leaving hens sitting for days without feeding, often brings its own punishment, as they become restless and unquiet and dissatisfied, frequently, in their desire to escape from their prison when no one is near, breaking some of the eggs. Considering the long period the hen sits, and the amount of care and patience she must, during this time manifest, it is only right and proper to make her as comfortable as possible. She will sit all the quieter when her appetite is satisfied, and she will be able to free herself from the small insects which are apt to be troublesome to hens, by the daily use of the dust-bath. These insects are particularly fatal to newly-hatched chickens, and it is always a shame to allow them to exist or increase when the means of preventing them are so easily attained.

In no case should a hen be fed on her nest, as such treatment, kind and gentle though it has sometimes been thought, almost always lames the hen, and prevents her from successfully brooding her chickens. Moreover, the circulation of a little fresh air is of great benefit in carrying off that stagnant vapour which proves so destructive to the life of the chick.

While the hens are absent, the eggs in the nest should be carefully examined. If the number falls short of the original sitting, search should be made in the nest for the missing egg, as very often an egg may be hid in the straw, or in a corner without its being noticed. If found it should be replaced. Not unfrequently it has been broken, and its contents scattered over all the other eggs. When this is discovered, the dirty straw must be immediately removed, the nest re-made, and the eggs severally washed in warm water, and then replaced in the nest. The breast of the hen should also be thoroughly examined, and, if necessary, washed before again putting her on the eggs. If this is not attended to, the result of the one broken egg will likely be that *few*, if any, of the remaining eggs will hatch at all.

It often happens that the hen lays after commencing

sitting, and when this is the case her eggs should be removed. So as to recognise them readily, it is well to mark with a circle of ink each of the eggs originally set.

At the end of the seventh or eighth day the poultry-keeper must examine all the eggs in each nest to see if they are fertile. It is easy to discover this, by holding the egg with the thumb and forefinger between the eye and a candle light. If, on the one hand, the eggs appear perfectly clear (as is the condition of new laid eggs), they are barren. If, on the other, the eggs are opaque or cloudy, they are fertile, and the chickens have begun to form. The eggs which continue clear are in no way destroyed by the few days sitting of the hen, and should therefore be kept for the use of the newly-hatched chickens. The fertile eggs must be divided amongst some of the sitting hens, and fresh settings should be placed under the remainder. If the eggs are not examined as I suggest, the hen continues to sit on perhaps many unfertile eggs, wasting her time and strength, when both might be turned to better account. After a few trials any one may be able to detect the barren eggs on the fifth day.

Watering the eggs daily is to be recommended, and the best plan is to use a flat camel hair-brush, dipping it in the early months in warm, and in the later months in tepid water, and then sprinkling the water over the eggs.

The day before the chickens are expected to come out all the eggs should be placed in a bowl of hot water—*no hotter than the hand can bear comfortably.* In the course of a few minutes the eggs containing living chickens will appear restless in the water and begin to move about—the eggs containing dead chickens will remain motionless or else sink. After leaving the whole soaking for about ten or twelve minutes, those indicating life should be again, while wet, placed under the hen. The dead ones should be removed. On the twenty-first day most, if not all, the chickens will be out from their shell, and the broken shells or fragments must then be removed at once. It sometimes happens, more especially during cold weather, that a day elapses after the twenty-first day, before all which have shown signs of life in the boiling water come out. But the poultry-keeper must not be disheartened at this, for even though he has to wait an extra day, he will find very few, if, indeed, any failure in their ultimate hatching.

If any considerable time elapses between the hatching of the several eggs, the chickens earliest hatched should be placed under one or two hens, and the remaining eggs placed under the other hens. This will prevent effectually newly-hatched chickens being crushed by the unhatched eggs, as they are not unfrequently apt to be.

The day the chickens are expected the hen should be well fed with hard food, so as to preclude the possibility of her becoming restless, in consequence of hunger, while they are hatching. Once they are all out they should be left alone and quiet with their mother for at least twenty-four hours. During this time they are perfectly independent of meat as they are full of the yoke which, immediately before hatching, passes through the navel into their bodies.

At the end of the twenty-four hours, hens and chickens

must be removed to their coops. If the coops are open they must, *especially in the early season*, be placed in a covered shed. Unused outhouses or offices will suit for this purpose very well, and if their flooring be either wood or stone, they must be covered with earth to a depth of *two* or *three* inches. The earth must be daily raked, and, if there is a constant succession of chickens filling the coops, renewed once a month. At the end of four days, or a week at the farthest, the coops must be placed on the grass. If a coop be employed for a hatching house such as I described and advocated in a previous chapter, one of its greatest advantages will be *the ease*, with which it may be converted into a chicken run. The straw must be removed from the hatching nest, and the hen cooped *there* instead. The yard, which before did good service as a feeding-yard for the hen, will now constitute a most excellent run for the chickens, well protected from wind and rain and all inclement weather. Coops must be often removed to fresh situations, as nothing is so baneful to the health of chickens, or so sure a progenitor of disease as tainted ground.

If there is an unlimited range, chickens should be allowed out in *very fine weather* for a short time daily. It will do them much good in bracing up their constitutions and in increasing their general stamina. Besides they are, even at this early age, eager and ravenous for slugs, worms, etc., and will cater for themselves and search out these insects with a quickness and readiness that is almost impossible to imagine, though most amusing to witness. During the first twenty days, and more especially during the first ten

GENERAL TREATMENT.

days of their existence, great care must be taken of them, as it is during that period they are most delicate. After that they will be in a great measure independent of so much constant attention.

At first both mothers and chickens should have all kinds of dainty food given—the former requiring it almost as much as the latter; but after a week only the chickens should be fed with it—the hens being supplied with ordinary hard and soft food three times a day. Hard boiled eggs and broken crumbs, mixed with milk, is, with the addition of a little meat, the best food for the chicken during the first few days of its life. During the second week the egg may be dropped and ordinary meals substituted for it. Nothing contributes more to their growth and early maturity than malt, and it should be given daily in the proportion of one handful to six chickens. Meal and rice and crushed bones should also be used. Patent meals, such as Spratt's, or Chamberlain's, or Thorley's food for cattle, are very strengthening, and give an agreeable variety to the feeding. Buck-wheat and hemp-seed must never be forgotten, more especially at night. Warm milk in the morning, and plenty of fresh water frequently renewed during the day, will be found most essential in keeping them in health. The chickens must be fed every *two hours* for the first week, *six or eight times* a day for the second, and afterwards not less than *four* times a day until they are three months old. The more attention they receive the better they will thrive. For six weeks the hens should be kept under the rips, unless they manifest a dislike to their chickens, when they must be immediately removed. Where this happens, and the chickens are still very young, they should in as large a flock as possible, be put together during the night in a *small house well littered with straw*. The natural heat of their bodies closely packed will keep them warm and comfortable.

If necessary, recourse may be had to an artificial mother, which may be constructed very simply out of a box not more than four inches high. It must have an open front, and be made to slant down towards the back. The inside should be lined with lamb's wool. Ventilation may be obtained by perforating the upper part of the ends with small holes.

If the largest profit is desired, capons should be trained to take charge of chickens—the hens themselves being sent back to their yards to lay, which they will recommence doing in about ten days or a fortnight. In France and other Continental countries it is no uncommon sight to see capons taking complete charge of innumerable broods of chickens. At first there is required both care and patience in training the capons for their maternal duties, but once trained they will continue during life the most careful of mothers, never apparently weary of their charge, and always ready to take care of fresh broods, however numerous.

One of the best ways to train a capon for his important duties is (having selected one that is good-sized, full-feathered, and about two years old) to confine him under a small dark box for about a week, care being taken that air is always admitted. Twice or thrice a day he should be fed on good stimulating food. At the end of a week, two or three young chickens, a fortnight old, should at the evening meal be allowed to feed with him, and by degrees permitted to run underneath his wings, which they will be sure to do if their own mother is absent. At first he may refuse to allow them, and may even ill-use them; in which case they must for that day be removed. By daily repeating the experiment he will soon welcome their presence as some kind of cheer amid his solitude, until at last he will actually become fond of them and very kind to them. After this he may be allowed his liberty; and having successfully taken care of the first brood, he will never seem satisfied unless in charge of another; and no hen could ever surpass, if equal him, in the attention and protection which he gives to his chickens. The weight of the chickens, both cockerels and pullets, is much increased by caponing, while at the same time additional delicacy is given to their flesh.

No real danger attends the operation if it is performed by experienced hands, and if due care is taken of the chicken both before and after. I do not, however, recommend any one attempting it who has not himself seen it done by another skilled at the work; for even with the most perfect information obtained from description or from reading, it can only be perfectly learned by personally witnessing the operation.

As Stephens in his "*Book of the Farm*" describes the operation very concisely, I shall here quote it. He says:—

"Capons of the common fowls are formed both of the cock and hen chickens when they are fit to leave the hen at about six weeks old. Chickens are transmuted into capons by destroying the testicles of the male and the ovaries of the female. The testicles are attached by a membrane to what is called the back-bone of the carved fowl. They are destroyed by laying the bird on its near side, keeping it down, removing a few feathers, and making an incision through the skin of the abdomen, and on introducing the forefinger through the incision, first the one, and then the other testicle is obliterated by pressure of the finger. In the case of the hen, the ovary is nipped off by the thumb-nail, or cut off by a knife. The incision is stitched up with thread, and little danger is apprehended of the result."

I need only add that it is necessary to keep the chickens without food for a day before, and to give them only soft food for a couple of days after, the operation is performed.

Constant variety in the food of chickens will keep up and quicken their appetites, which are apt to flag by long-continued feeding on the same kind of food. A little citrate of iron, or some of what I have previously recommended, the Douglas' Mixture, added to the water every third day, will brace up the constitution of the chick in a wonderful way, and prevent the same liability to disease in wet weather. Ale may also with much advantage be mixed with the food. Such cockerels as are intended to be retained until grown up, *must be separated* from the pullets when *12 weeks old.* Those not thus intended must be killed, at the same age. Before killing, they must be kept in a fattening coop, for a fortnight, and fed unsparingly on the most flesh-forming kind of food, such as Indian meal and buckwheat meal. When this food is mixed with milk, it gives a beautiful white appearance to the flesh. Immediately after the cockerel is killed, it should be plucked, and then for an instant dipped in boiling water. It should be killed the day it is to be used.

The pullets, when separated from the cockerels, must be fed on the most stimulating food, so as to induce early laying. Bright redness in their combs indicate that the fowls, whether pullets or hens, *are laying*.

CHAPTER III.

DAILY DUTIES.

THE daily duties connected with a poultry-yard must be discharged with the utmost regularity. Once their details have been thoroughly mastered and perfectly understood, they will not occupy many minutes each day. Personal supervision will do more than almost anything else to contribute to the success of the establishment, be it large or small, and should not therefore be neglected.

The first duty, each day, is to visit the hen-house or houses, the yard or yards, and to examine the feeding-troughs, the drinking vessels, and the nests—*not one of them* must escape minute and close scrutiny, and if not perfectly clean must *at once* be made so. It will then be necessary to see that lime and sand are supplied, for without the former, the shell will not be formed, without the latter, the food will not be digested. In wet, or very cold weather, or during moulting, a little citrate of iron, or some of Douglas' Mixture should be added to the drinking water, just sufficient to give it a bitter taste. Once a week lime water should be given.

While the inspection of the houses, etc., must not be forgotten it is as essential, if not more so, to see that every hen is in good health. At first it may be rather difficult for the amateur to say what is specially *the matter* with a hen apparently unwell, but there is no difficulty in distinguishing between a hen in good, and in bad health. Whenever it is seen moping in a corner, with its feathers ruffled and its wings drooping, there is at once an indication that it is ill. It should be immediately separated from the others, as it is apt to communicate disease to them, and once disease breaks out in a yard, it is not easy to avert its progress. If the day is wet, the hens must be confined to their covered yards, in which the dust-bath should always be kept. Once a fortnight a little black sulphur should be mixed with it.

It must be seen that the feeding *has been* and *is being attended to* with perfect regularity. Never later than six in summer, and always at daylight in winter, the first meal should be given. It should invariably consist of soft food. Twice a week variety may be given to it by substituting one meal for another in the preparing of it. This will serve to revive and stimulate the most flagging appetite.

Never more than twenty-five fowls should be kept in one yard, as the difficulties of feeding them properly, are materially increased by an increased number. The weaker fowl is sure to get less than it required, through the rapidity with which the strong and healthy eat. The attendant should take plenty of time in feeding, and endeavour to regulate the amount which each fowl gets.

The food is always better when given warm. It should therefore be mixed with hot water, or, what is better, boiled the night previous to being used, and left beside the fire all night. When a special boiler is used after the food is boiled, the fire should be drawn, and it will be found in the morning that the food has retained the requisite degree of warmth for use. With the exception of summer, when fowls can find insects for themselves, it is absolutely necessary to give them animal food daily, and if they are penned up the whole year, it should not be discontinued even in summer. If the table refuse is not sufficient, bullock's liver or sheep's pluck may be purchased at a cheap rate. A very strong, though a *very foolish and stupid prejudice* exists against the use of horse-flesh. Many a time the carcass of a horse killed by accident, or of an old one destroyed, might be purchased at a very small cost, and better, or more stimulating food could not be given to poultry. If used, it should be boiled, and, like all other animal food, be minced and mixed up with the morning diet.

It may perhaps be useless to refer to a way adopted in France, for the germinating and constant supply of insect life, as one not likely to commend itself to our national tastes, but in case that *even a few* of my readers may wish to know, and try for themselves, the plan so successfully carried out on the Continent, I shall here explain it. A pit is dug to a considerable depth, and then filled in with alternate layers of straw chopped fine, horse dung, vegetable soil, flesh of dead animals, offal, etc., etc. To prevent rain destroying or fowls eating the contents, it is necessary to have the pit whenever filled well covered. In a short time the whole mass will be alive with maggots, worms, and insects which may be given to the fowls. The whole thing may be constructed most cheaply, and however much poultry-keepers may object to have their poultry fed on it, the poultry have no objection to partaking of it, as may be witnessed in the eagerness with which they consume it. Indeed, when one thinks how necessary insect life is to the proper sustenance of fowls, it seems most unreasonable to object to supply to poultry by artificial means, what nature has amply provided for them in their natural state. When given judiciously it will materially increase the daily supply of eggs.

Tallow chandler's greaves or cracknel cheese is an excellent substitute for meal, and is of much importance in enriching the food of poultry.

The food of laying hens should daily be seasoned with cayenne pepper, ginger, or mustard. This will tend to increase their laying powers. Salt should also be given, in no larger proportion, however, than is used in ordinary household cooking.

So much then for the morning meal. At twelve o'clock, when fowls are kept constantly in confinement, a *very scanty supply* of whole grain should be given to them, and again at six in summer, or before dark in winter, the last feed for the day should be given.

I recommend that the last meal should consist of boiled barley, boiled wheat, oats, or buck-wheat. The substitution of one grain for another twice a week will be found advantageous, as thus variety is given to their evening as well as to their morning diet. If the fowls are fed at mid-day the quantity given at night must be very small. When fowls have complete liberty, *two meals a day* will be sufficient; and if such is the plan adopted, more must be given at the last feeding than is necessary when a twelve o'clock meal is supplied. The great advantage of grain in the evening is, that it takes a much longer time to digest than soft food, and affords both support and warmth to the fowls during the long night.

Before passing from the subject of feeding, I may here mention, that the proper mixing of soft food is of very great importance. The thin porridgy mess that I have sometimes seen provided is not only a source of annoyance to the fowls, as it sticks on their beaks, but is very apt to bring on diarrhœa. Again, the dry, mealy mixture, that when given, resolves itself into a kind of powder, is equally objectionable. At all times the food should be worked up into such a consistency that when thrown on the ground it will break.

If possible, it is much better to avoid the use of a feeding dish or trough—though this cannot always be done. Those who can feed on a grassy sward, or whose yards are dry and clean, should scatter the food over it; for not only will the fowls take a longer time to feed, but they will also pick up grass and sand, which are beneficial in assisting digestion. For small yards, in wet weather a dish must be provided, and the best kind I can recommend are open ones made of iron, with bars of iron across to prevent the fowls treading on or destroying the food.

The knowledge of the proper quantity to be given daily is of immense importance to the poultry-keeper. Perhaps one of the best rules to be observed is to *cease feeding* whenever the birds cease to run after the food or seem careless about it. As much waste generally takes place in the preparing, as well as in the giving of food, I strongly recommend that the quantity required for each meal should be measured beforehand, and never *exceeded for any reason whatever*. A *small handful of grain*, or a *tablespoonful* of *soft food* is an *ample full feed for each* fowl of the larger breeds—a *less quantity* will do for the *smaller*. The exact number of hens kept *should always be accurately known*, and in no case should the supply for each be greater than I have prescribed. They will always be ready for their food under this treatment, and be in excellent health, while the cocks will be more vigorous and the hens better layers.

When a fowl is being fattened for table use, it should, *for a fortnight* before being killed, be cooped in a small pen, of sufficient size to accommodate it, without allowing it room to turn. The coop should have a barred floor to admit of the excrements falling through, and a barred front to enable the fowls to feed. The food should be soft and of the most fattening kind, such as buckwheat meal, Indian meal, rice, and milk, and should be given *abundantly, ungrudgingly*, above all, *punctually*, four times each day in a semi-liquid state. Water need not, and grain *must not*, be allowed.

The cramming system so generally adopted in France is not at all likely to find much favour in this country, so that it is unnecessary to describe it.

At the end of a fortnight, when the fowl is sufficiently fattened for use, it should remain for *twelve* hours without food, and then be killed either by the insertion of a sharp knife in the roof of the mouth, or by very quickly dislocating the neck. If the former plan is adopted, the bird must be hung up by the legs immediately after death, so that it may bleed freely, otherwise the blood is apt to pass through the body, and to prevent that whiteness of flesh which is so desirable.

Immediately after plucking the feathers, *which should be done while the bird is warm*, the carcass should be dipped into boiling water for a few seconds. Fowls should either be used the day they are killed, or kept for several days. In the case of old birds the latter plan is the best, and when used they should be parboiled and afterwards roasted. In this way they are made more tender, for as a general rule they are very tough, and unless some plan is adopted to render them less so, they are not worth eating. It is to be hoped, however, after what has been said, that there will be no old fowls kept at all.

CHAPTER IV.

PROFITS.

I HAVE now given in detail all the information that is requisite for the thorough and perfect management of poultry. I can promise *success* if only the many directions I have furnished are strictly adhered to. I simply ask a fair trial of the *régime* unfolded and explained in these pages, and I am certain that even the most sceptical will be induced to believe that poultry may be made to pay, and to pay handsomely.

I propose to draw attention to the profits which may with great ease be realized by the judicious management of poultry. I promised in the outset of my treatise to show how eggs at *fourpence per dozen* and chickens at *fourpence per pound* can be obtained, and I shall now prove the correctness and truth of my assertion.

I have already spoken of the *exact quantity of food which should be given* to each fowl. If this quantity is *strictly adhered to*, the price of feeding will never exceed *one penny per week*, averaging the winter against the summer. I have no doubt many who have never given the subject a thought will be prepared to sneer at this declaration; but I can assure them, that *the very best authorities* on poultry management, have proved that one penny per week is quite ample to keep fowls in highest health and best laying condition. When they cost more, it may be safely concluded that there is considerable waste of food, or that more is given than is necessary. I have, after considerable experience, established the fact for myself, and I would only say to any doubters "*try for yourselves,*" and if they give my recommendations a fair trial, preventing all waste, I am convinced they will at once cease to be unbelievers.

The average number of eggs laid in one year by the common barn-door hen seldom exceeds 90 or 100. The pure breeds I have been recommending lay double that number. From the best sources, it has been proved that Houdans will average 200 eggs from each bird in the year, Andalusians and Leghorns 220, and Hamburgs 220 to 240. From this estimate, *which may be relied on*, it can easily be seen what an *immense advantage* it is to the poultry-keeper to *keep only* these valuable breeds. Now, as I do not wish to be extravagant in my estimate, let us take the very lowest average of eggs (180) produced by a single hen of the fancy breeds in a year. This gives 15 *dozen* from one hen—a by no means *too large* quantity to expect. Taking into reckoning the average price of feeding the hen at one penny per week (4s. 4d. per year), the price of each dozen of eggs would be, on the nearest calculation, 3½d.—*one halfpenny less* than the amount I promised. Suppose, however, as may easily be the case, 200 eggs per year is the quantity

PROFITS.

supplied, the price of each dozen would be rather less than 3d.

It may perhaps be urged that in this estimate I have not taken into account the cost of the hen, from the time of its hatching until it has begun to lay. In reply, I would say that *there is* to *shew against the expenditure* in rearing the chicken, the hen itself, which, at the age of eighteen months, should be worth at least 2s. 6d. The cost of rearing the chicken, even at the dearest estimate, will not exceed 2s. 2d.

Then as to the production of meat, a chicken, for the first two months of its existence, will not cost on an average more than one halfpenny per week, when fed on ordinary food; but, as I have recommended more *expensive food* as the very best means of hastening it through its tender age and hurrying it to maturity, I shall allow for it, until it is three months old, the same sum of one penny per week as I have allowed for an adult hen. This will easily cover all the extra expense of the higher feeding. For two weeks before killing it will be necessary to give the chicken extra fattening food, at a further cost of say threepence per week, which will bring the full sum up to 1s. 6d. when the bird is three and a half months old. Its weight at that age, if the breed has been judiciously selected, will be at least *five pounds*. Taking the whole cost at 1s. 6d., a chicken has been obtained at, as nearly as possible, 3½d. per pound—again *one halfpenny less* than the amount I promised.

Before passing from this part of my estimate, I must add, that it is perfectly possible for householders to obtain both eggs and chickens at *a cheaper rate* than I have mentioned. No one who has had charge of a house for any time, but will admit that there are innumerable scraps of food, in the shape of meat, puddings, crumbs of bread, potato peelings, etc., etc., which are thrown daily into the ash-box. Now, if poultry are kept, all the refuse of the house, which is otherwise wasted, will be utilized, and thus at a very trifling cost, eggs and flesh may be regularly supplied for the table.

So much, then, for mere household supply. I propose to look at poultry from another point of view, viz., the cotter's and farmer's who rear them for sale and profit.

The cost of a fowl from the time it is hatched to the date of its beginning to moult at eighteen months old is, including the original cost of the egg (at say one penny), 6s. 7d. To that let us add 6d. for a fortnight's fattening, and 5d. for any incidental expenses. The whole expense of a single fowl, for the eighteen months of its existence, amounts thus to 7s. 6d. So much then for the expense. The return will be made in eggs, in the excrements, and, finally, in the *sale* of the fowl—all of which will realize at least 19s. 6d. ; from which, if the cost, 7s. 6d., is deducted, there will remain 12s. of profit. Now, in this estimate, I am on the one hand giving the original price of an egg for hatching, the cost of a fowl's keep and of its fattening at *their highest rate*, and on the other I am giving the number of eggs laid in a single year, much fewer than may be expected, and the price of the eggs and of the hens when sold at a cheaper rate than they will realize, and the excrements at a lower value than they will fetch. I daresay some people may be inclined to dispute my right to include the manure of the fowl as part of the profit; but just as the profit in the keeping of cattle is often found to be materially increased by the manure produced, so one very important item in poultry keeping is the excrements of the birds. More valuable manure, and more especially for market gardeners, it is impossible to obtain; and when carefully collected and kept dry in old casks, mixed with earth (for it is too strong to be used alone), it is equal in its properties to guano. It is almost inconceivable the quantity of manure that a single bird will void in a year. Three or four ounces are a common quantity to be dropped in one night by each bird of, at any rate, the larger breeds. The value of each cwt. of poultry manure is, *according to the best authorities*, 7s. Now, if any one cares to calculate the quantity which each fowl will void in a year, at the rate of two ounces per night, and at the price at which it is ordinarily valued, the estimate of 2s. per annum for the excrements of each fowl, will be found to be below rather than above the right value. To sum up all in a few figures, the Dr. and Cr. account will stand thus :—

Dr.		Cr.	
Cost of setting egg,	£0 0 1	180 eggs at 1s. per dozen,	£0 15 0
Cost of feeding hen from time of hatching till 18 months old,	0 6 6	Price of fowl when sold,	0 2 6
		Excrements,	0 2 0
Cost of fattening,	0 0 6		
Incidental expenses,	0 0 5		
			£0 19 6
		Deduct expenses,	0 7 6
	£0 7 6	Total profit,	£0 12 0

The eggs I have only valued at 1s. per dozen, a very low average considering the high prices which are now paid for them in towns and centres of industry. In winter they are seldom if ever to be purchased anywhere under 2s. per dozen, and in summer, when they are cheapest, I have never found the least difficulty in getting 1s. per dozen for them. It may be easily seen how very much the profit would increase if, as is not unlikely, a higher price is obtained than I have named, or if more eggs than the 180 are laid. If the breed is well chosen, and the management carefully attended to, 200 eggs and more will be laid by a single hen, more especially if, when hatching, only the eggs are set of prolific layers (as, in every breed, some hens unquestionably are).

In this way it is quite possible to have in a few years a race of hens that will lay nearly 300 eggs per year each. I need not remind the reader that the offspring of cows celebrated for their milking qualities, generally inherit these good qualities, and even yield a larger quantity of milk than their dams ever did. And so, if a hen noted for its laying be mated with a cock sprung from a hen equally prolific, the laying powers of their offspring will be largely increased.

I have only allowed two shillings and sixpence as the price of the hen when sold at the age of eighteen months. That is the price of *common barn-door fowls* NOW. Whereas if the fowls I have been recommending are adopted, the size of each fowl will be much greater, weighing eight or ten pounds, and the flesh more tender and nutritious. Eightpence a pound would not be too much to ask or expect, and taking the weight of the fowl at eight pounds, five and fourpence would be the price obtained. To this the objection may be offered, that it is impossible to sell fowls by the weight. Perhaps in the present state of things, such is the case, for no vendor cares to risk weighing the miserable deformed creatures now sold as good fowls, or how fearfully would they be found wanting. But depend upon it, if *large, plump, meaty, tender* fowls, weighing eight or ten pounds were brought into the market, they would soon command the price I name. And that there is already a promise of better times, both for poultry and the poultry-keeper, I think I need only mention the fact that capons and turkeys are being sold by the pound in the large English towns, and that the former bring as much as seven and eight shillings each.

Once more I must draw attention to the profits which every hen is capable of returning. In the outset of my treatise, I remarked that a profit of ten shillings is what might be expected, while in my Dr. and Cr. account I have made the actual profit twelve shillings. The odd two shillings I leave to cover expenses of houses, attendance, etc., and if each fowl be charged at the rate of two shillings per annum for accommodation and attendance, it certainly pays most handsomely, if not extravagantly. If any saving is effected, as certainly there may be, on these items, the profit will be proportionately increased.

So far I have directed attention to the egg production as the most profitable branch of poultry-keeping. I will now allude to the rearing of chickens for the market. The great thing in this case is to select *that* breed which arrives *quickest at maturity* and *fattens most readily*. *One thing* must be determined on, that the chickens will *only be sold by the pound weight*. Where eggs are known to be fresh, a large sale can readily be secured; and when it is known where *good, well-flavoured, tender, weighty chickens* may be obtained, customers will soon become numerous; and above all things, pay the price that is asked, if they feel assured the quality is excellent and guaranteed. The following may be taken as the Dr. and Cr. account:—

Dr.		Cr.	
Cost of egg,	£0 0 1	Value of chickens at 8d. per pound—5 pounds,	£0 3 4
Cost of feeding fourteen weeks,	0 1 2	Excrements,	0 0 4
Cost of fattening,	0 0 6		£0 3 8
Incidental expenses,	0 0 5	Deduct expenses,	0 2 2
	£0 2 2	Total profit,	£0 1 6

This branch of poultry-keeping is not so profitable as the other (the egg production), but it has this advantage, that one's capital is turned over several times in the year, and that the quick return enables the poultry-keeper of limited means the more readily to extend his establishment.

As will be seen, I have valued the chickens at their ordinary market value, without reckoning the extravagant prices which are at times paid for them. Spring chickens will fetch as much as 5s. and 6s. each in London early in the season, while often, notably during the Epsom week, as much as 10s. has been given for one. Of course these high prices are exceptional, but it is always well for the poultry-keeper to keep special seasons in view, and on the principle that it is best to make hay when the sun shines, to have ready for market what he can dispose of, at very considerable profit.

I must also mention that in the foregoing accounts I have given 1d. as the price of the egg for setting. Now, though this might be reckoned as the price of each egg, as I have previously remarked, when poultry-keeping is commenced, they should not exceed at the most 4d. per dozen after the whole establishment has been got into working order.

THE END.